Cambridge Elements

Elements of Paleontology

THE NEOTOMA PALEOECOLOGY DATABASE

A Research-Outreach Nexus

Simon James Goring
University of Wisconsin-Madison

Russell Graham
Pennsylvania State University

Shane Loeffler
University of Minnesota

Amy Myrbo
University of Minnesota

James S. Oliver
Pennsylvania State University

Carol J. Ormand
Carleton College

John W. Williams
University of Wisconsin-Madison

CAMBRIDGE
UNIVERSITY PRESS

University Printing House, Cambridge CB2 8BS, United Kingdom

One Liberty Plaza, 20th Floor, New York, NY 10006, USA

477 Williamstown Road, Port Melbourne, VIC 3207, Australia

314–321, 3rd Floor, Plot 3, Splendor Forum, Jasola District Centre,
New Delhi – 110025, India

79 Anson Road, #06–04/06, Singapore 079906

Cambridge University Press is part of the University of Cambridge.

It furthers the University's mission by disseminating knowledge in the pursuit of education, learning, and research at the highest international levels of excellence.

www.cambridge.org
Information on this title: www.cambridge.org/9781108717885
DOI: 10.1017/9781108681582

First published 2018

A catalogue record for this publication is available from the British Library.

ISBN 978-1-108-71788-5 Paperback
ISSN 2517-780X (online)
ISSN 2517-7796 (print)

The Neotoma Paleoecology Database

A Research-Outreach Nexus

Elements of Paleontology

DOI: 10.1017/9781108681582
First published online: October 2018

Simon James Goring
University of Wisconsin-Madison

Russell Graham
Pennsylvania State University

Shane Loeffler
University of Minnesota

Amy Myrbo
University of Minnesota

James S. Oliver
Pennsylvania State University

Carol J. Ormand
Carleton College

John W. Williams
University of Wisconsin-Madison

Abstract: Paleoecological data from the Quaternary Period (2.6 million years ago to present) provides an opportunity for educational outreach for the Earth and biological sciences. Paleoecology data repositories serve as technical hubs and focal points within their disciplinary communities and so are uniquely situated to help produce teaching modules and engagement resources.

The Neotoma Paleoecology Database provides support to educators from primary schools to graduate students. In collaboration with pedagogical experts, the Neotoma Paleoecology Database team has developed teaching modules and model workflows. Early education is centered on discovery; higher-level educational tools focus on illustrating best practices for technical tasks.

Collaborations among pedagogic experts, technical experts, and data stewards, centered around data resources such as Neotoma, provide an important role within research communities, and an important service to society, supporting best practices, translating current research advances to interested audiences, and communicating the importance of individual research disciplines.

Keywords: paleoecology, curriculum, outreach, engagement, databases, community curated data resources, climate change, geologic time, global environmental change, paleoecology, proxy data, teaching, engagement, Neotoma Paleoecology Database

ISBNs: 9781108717885 (PB), 9781108681582 (OC)
ISSNs: 2517-780X (online), 2517–7796 (print)

Contents

1 Introduction 1

2 The Neotoma Paleoecology Database 3

3 Neotoma Explorer: Data Discovery and Exploration 5

4 More Exploration with Explorer: Additional SERC Teaching Exercises 9

5 Educational Modules for Climate Change, Paleoecology, and Biogeography 9

6 Programmatic Access for Macro-scale Paleoecological Research (APIs, neotoma R, and GitHub) 11

7 Flyover Country®: Beyond the Classroom 13

8 Discussion 14

References 18

1 Introduction

The Quaternary Period (2.6 million years ago to present) is a highly active area of integrative ecological, geological, and climatic research. Quaternary timescales range from the decadal and centennial scales of our lives and historical memory to deep-time geological timescales of millions of years and longer. In addition, the Quaternary has been a time of planetary-scale climate changes, operating at many timescales, recorded in many different types of datasets, and with strong and clear effects on cryogenic, biological, and physical systems. This combination of factors – the range of timescales including human history plus the data-rich record of planetary climate change – makes the Quaternary a natural entry point for teaching and learning about paleoecology, geologic time, and climate change via authentic data exploration (e.g., Manduca and Mogk, 2002; Hunter et al., 2007; Laursen et al., 2010; Resnick et al., 2011, 2012).

The Quaternary is of particular interest to scientists studying climate change (Masson-Delmotte et al., 2013), the ecological effects of climate change (Araújo et al., 2008), and the causes and effects of species range shifts and extinctions (Ordonez and Svenning, 2017), providing engaging questions for students. Climatically, the defining features of the Quaternary include: (1) the oscillation between long glacial periods and short interglacials, accompanied by the growth and collapse of continental ice sheets, sea-level falls and rises, and variations in global temperature and atmospheric greenhouse gas concentrations; (2) a pacing of these glacial-interglacial changes by variations in the Earth's orbit, known as Milankovitch cycles; and (3) large and abrupt temperature changes during glacial terminations such as Heinrich Events and Dansgaard-Oeschger events (Masson-Delmotte et al., 2013). During the Holocene interglacial (11,700 years ago to present), temperature variations have been muted but hydrological variability has intensified (Mayewski et al., 2004), linked to both internal processes such as shifts in ENSO variability and external forcings such as individual volcanic eruptions and variations in solar luminosity (Wanner et al., 2008).

Ecologically, species distributions repeatedly retracted and expanded during the glacial-interglacial cycles (Davis and Shaw, 2001), with high local community turnover during climatic events (Blois et al., 2010). The worldwide expansion of humans before and during the last deglaciation was accompanied by a global wave of extinctions of large animals (Pleistocene megafauna) and other hominid species (e.g., Neanderthals, Denisovians). During the Holocene, post-glacial range expansions of tree species continued, with population dynamics and range expansions often paced by hydrological variability and

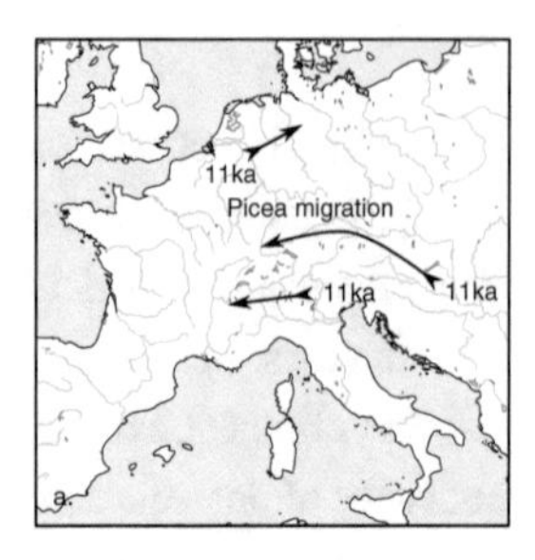

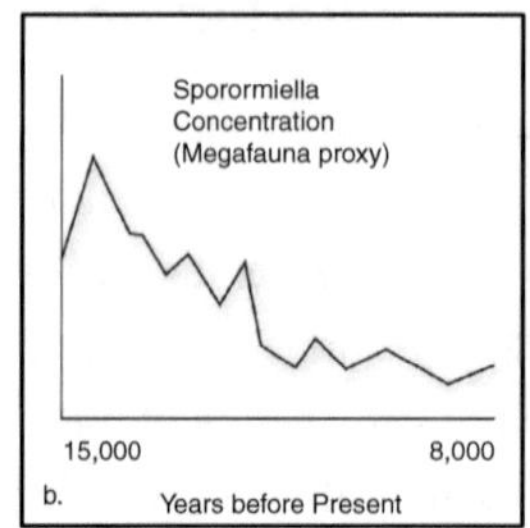

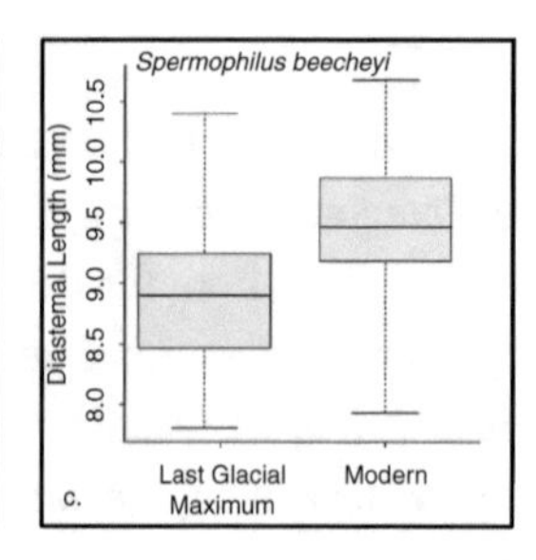

Figure 1. Examples of Quaternary datasets, simplified and adapted from their original publications: (1) Species migration through time beginning 11,000 years before present (11 ka), here, *Picea* (spruce) in Europe after the end of the last ice age (from Latalowa and van der Knaap, 2006); (2) Megafaunal extinction interpreted from the decline in the fungal spore *Sporormiella*, present in the dung of large herbivores (after Gill et al., 2009); and (3) change in mammal body size in the ground squirrel *Spermophilus beecheyi* (after Blois et al., 2008).

drought. Understanding these past ecological dynamics helps global change ecologists and biogeographers understand how species are likely to respond to current environmental changes and test the predictive ability of Earth system models.

Our understanding of climatic and ecological dynamics during the Quaternary is founded upon many individual site-level analyses of fossil organisms and paleoenvironmental proxies (Fig. 1). Fossil data are many and varied and include micropaleontological data (e.g., pollen, diatoms, ostracodes, foraminifera), macropaleontological data (e.g., vertebrate fossils, plant macrofossils), and, increasingly, molecular and organic geochemical tracers such as ancient DNA or organic compounds such as alkanes from leaf waxes. These paleoenvironmental records are collected from a diverse array of depositional environments, including excavations, lake sediment core sampling, or other fossil localities. Understanding the large-scale spatial phenomena that define the Quaternary requires gathering these many diverse datasets into larger databases that store information about age, taxonomy, spatial coordinates, depth or stratigraphy, and other related information (Williams et al., 2018).

In response to these needs, the Neotoma Paleoecology Database (Neotoma) has emerged as a community-curated data resource (CCDR) with the mission of gathering, curating, and sharing paleoecological and paleoenvironmental data, to enable open, global-scale science (Williams et al., 2018). This resource also provides students and educators with a public portal for accessing data that can be used to understand ecological change through the last several million years. Authentic inquiry into real, complex datasets engages students in the learning process, thus deepening their understanding of species dynamics in changing environments, and simultaneously increasing their retention of key concepts (e.g., Laursen et al., 2010; Lopatto, 2010; Freeman et al., 2014). In recent years, new systems have been developed for finding, visualizing, exploring, and obtaining Neotoma data (Williams et al., 2018). Several recent educational and outreach activities have resulted in the production of a number of new resources for a wide range of audiences, including high school and college students, graduate students, and the general public (https://serc.carleton.edu/neotoma/index.html). Phone-based apps have been developed for travelers to understand the geological and ecological history of the world around them (Flyover Country®: http://flyovercountry.io).

This chapter reviews these resources and introduces Neotoma as a new platform for supporting authentic teaching and outreach, able to support a wide range of users and levels of expertise. We begin with an overview of the Neotoma database and its software ecosystem, then move to an introductory exploration of Neotoma's data holdings, using the map-based Neotoma Explorer and a variant of exercises developed with the Science Education Resource Center (SERC) at Carleton College (https://serc.carleton.edu/index.html). We then summarize teaching exercises for upper-level undergraduate and graduate students using the software package R (Goring et al., 2015) that have been developed through a series of workshops. Next, we show how the Flyover Country® mobile app can be used to discover and learn about Earth's history, drawing on and integrating resources at Neotoma and elsewhere. We conclude this chapter by examining the role of community-curated data resources as bridges between the research community and the public, with regard to educational outreach in particular.

2 The Neotoma Paleoecology Database

Neotoma comprises both a data repository (Neotoma DB) and an ecosystem of interlinked software for accessing and visualizing Neotoma data. Neotoma DB stores more than 25,700 unique datasets from 12,500 distinct sites, each containing observations of fossil organisms and associated geochronological and paleoenvironmental information spanning the last several million years from continental sites, including fossil beds, lakes, mires, and other depositional environments. Neotoma DB is built around a flexible data model

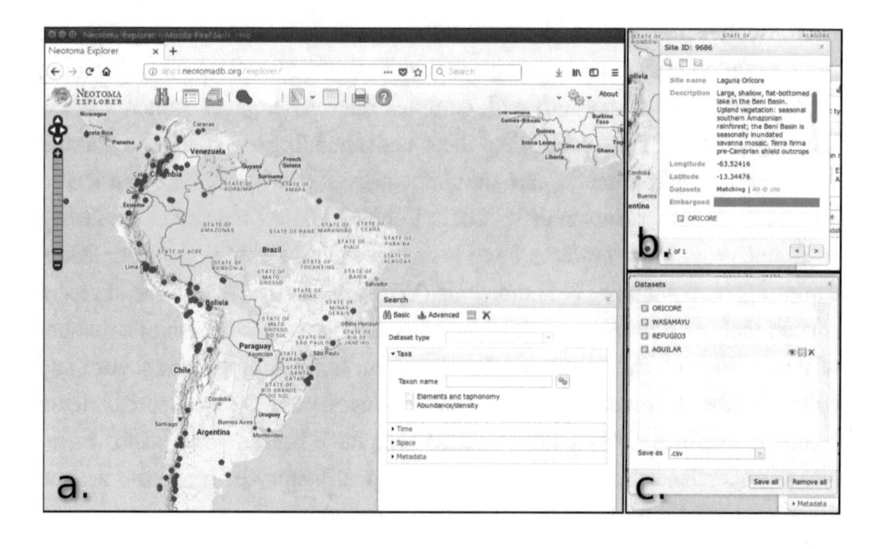

Figure 2. The Neotoma Explorer is a web-based data discovery tool that allows a map-based search for paleoecological sites using species names, site locations, site names, or other associated metadata (a). The tool provides the opportunity to examine individual sites (b) and to select a number of sites for download and later processing or study (c).

designed to describe paleoecological data records for many types of sites, depositional environments, taxonomic groups, and geochemical and sedimentological measurements (Grimm, 2008; Williams et al., 2018). The Neotoma DB includes information about researchers, locations, publications, time, and the raw paleoecological and paleoenvironmental records that underpin our understanding of Quaternary environments. The database includes site-level information for many of the datasets that includes a narrative description of the site, and can be linked to environmental data including water chemistry, regional climate, and other data based on spatial relationships.

Neotoma Explorer (https://apps.neotomadb.org/explorer/) is the primary data portal and discovery interface for finding and exploring paleoecological records within Neotoma (Fig. 2). Neotoma Explorer relies on a map-based user interface, in which users can search for data by taxonomic name, site name, spatial domain, time range, or many other dimensions. Discovered data are shown as sites on maps, which users can click to discover more information and download data. Neotoma Explorer is powered by an application program interface (API; http://api.neotomadb.org) that can be separately and directly accessed through any standard browser or programmatically. The use of an API provides a mechanism to build new tools on top of Neotoma's data resources. The API also supports the neotoma

R package (Goring et al., 2015) and the inclusion of Neotoma data within the Flyover Country® mobile app, and search tools that can jointly search Neotoma and other paleobiological resources (http://earthlifeconsortium.org/).

3 Neotoma Explorer: Data Discovery and Exploration

The following introductory exercise is based on Neotoma Explorer and is designed to help individuals understand how paleoecologists interpret and process proxy data, ranging from a single site to continental scales and from the recent past to deep time records. The exercise, based on "Exploring the Neotoma Paleoecology Database" (https://serc.carleton.edu/neotoma/activities/121251.html), is directed toward college undergraduates and explores changes in mammal and vegetation distributions since the Last Glacial Maximum. This exercise, along with others, was developed in 2015 at a workshop hosted by SERC. The workshop provided an opportunity for paleoecologists to work with science education researchers to develop high-quality teaching materials for a variety of educational levels. Additional educational modules are available on the SERC-Neotoma webpage (https://serc.carleton.edu/neotoma/activities.html) and are described further below.

"Exploring Neotoma" teaches users how to search for paleoecological data through a variety of starting points (by site name, dataset type, and taxon name), find publications, and create simple mapped visualizations that show (1) how species ranges changed as climates changed and ice sheets retreated and (2) changes in associations among plant and animal taxa. The exercise introduces students to research questions that paleoecologists might ask; for example, questions about species co-occurrence and shifting distributions under changing climate scenarios. "Exploring Neotoma" engages students through the authentic exploration of paleoecological datasets, and provides a foundation of skills they can use for further exploration and inquiry in subsequent exercises. The following sections, through to Research Questions, are cumulative and are expected to follow one another.

3.1 Searching for Data with Neotoma Explorer: A Student-Centered Walkthrough

The Neotoma Explorer provides the spatial overview and search capabilities for Neotoma. One of the simplest things you can do with the database is to search for a single research site. To do this, look for the *Metadata* search option under the *Advanced* tab of the search panel (Fig. 3). Marion Lake is an important paleoecological record from western North America, which provided one of the first quantitative reconstructions of Holocene climate from

Table 1 Data tabs presented in the data view of the Neotoma Explorer for selected datasets. Data tabs are labeled in the dataset view. Information presented in each tab is described here.

Data Tab	Table Description
Samples	A table showing individual pollen counts, depth information, and summary chronology
Diagram	A tool to view the change in the changes in taxon presence or abundance over depth or time
Site	Site-level information for the data
Chronology	The age-depth model used at that site and dataset
Publications	Publications related to the dataset

Figure 3. Searching for Marion Lake using the Neotoma Explorer's search tools (https://apps.neotomadb.org/explorer/).

western North America (Mathewes and Heusser, 1981). Search for Marion Lake under the *Site Name*. You may find more than one site. Which one is the pollen dataset? How do you know?

The northernmost site is the Marion Lake pollen sample site. When you click on the point, you should see some site information and a description, along with a green *P*, as well as a small clock (*clock*) icon. Both have the word *MARION* beside them. The icons represent pollen data and geochronological data respectively. Click on the icon that represents the pollen data (the green *P*) first and scroll through the tabs available within the window (Table 1). You'll see the following:

These window tabs provide you with a pretty good overview of the dataset, long-term changes at the site, and a list of the publications available to learn more about the sample record.

3.2 Searching by Dataset Type

Neotoma can also be used to look for a large number of datasets within a particular dataset type. Create a new search (make sure you clear the site name). Next to *Dataset Type*, click the down arrow to display a list of choices. What do you see? Select *water chemistry datasets*, perform the search, and then click on a site to view the data, as you did for Marion Lake. How do the variables in the water chemistry diagram differ from those in the pollen diagram? What similarities in the data can you see? Are rates of change different? What might be driving differences or similarities between these records?

3.3 Searching by Taxon Name and Time

One very quick, informative search in Neotoma Explorer is to look for a particular species or taxon, within a set time range. (You can also restrict searches spatially; Explorer's default is to look within the bounds set by the mapped view on your computer.) Let's look for fossil pollen sites with *Picea* (spruce) present between 15,000 and 12,000 years ago. This time period is of interest because it is right at the end of the Pleistocene, as ice sheets are retreating, humans are arriving and dispersing across the Americas, and plant species ranges are shifting north. Go to the *Advanced* tab, make sure all prior searches are cleared, then select the term "pollen" in the *Dataset* type field and "*Picea*" in the *Taxon Name* field. Then check *Abundance/density* and set abundance to ">20%" (to only show sites with a lot of *Picea* pollen at them). Last but not least, click on the *Time* bar and enter a time range of "15,000 – 12,000" ybp (years before present).

3.4 Mapping Ice Sheets

After the last search you should have a map with dots showing all the fossil pollen sites with at least 20 percent spruce pollen between 15,000 and 12,000 years ago. Feel free to explore them further. But where were the ice sheets? Click on the *white polygon* on the top bar to display glacial boundaries. This supports the addition of overlays to map glacier extents through time. Enter a date (say 12800 ybp), hit *enter*, and then *zoom* out the map so it shows most or all of North America. You should now see a white polygon, representing the continental ice sheets, overlain on the map of sites. You may also see blue polygons, which show the large lakes (proglacial lakes) that formed around the glacial margins, fed by meltwater. If you still have the spruce (*Picea*) sites displayed, you should see that spruce was growing just south of the ice sheets. If you move the time window for the spruce search to a more recent date, you should see that the site distribution has moved northwards, which represents migration in response to warming climate.

3.5 Research Questions

We all know that a whole bestiary of large animals roamed the Earth during the last age, many of which are now extinct. In North America, representative species of the Pleistocene megafauna include mastodons (*Mammut*), mammoths (*Mammuthus*), saber-toothed tigers (*Smilodon*), and many others. Where did these animals live, did their distributions change over time, and were they associated with particular habitats? We have already mapped the spruce trees and ice sheets, so now let's see where the big animals were. Let's use mastodons in this example. Do a taxon search (making sure that "Mammals" is entered under *Taxa Group*) for *Mammut* (mastodons) and select all *Mammut* taxa using the *Multi Taxon Search* tool, presented as a *Set of Gears* beside the *Taxon name* field (Fig. 4). Look at the site distribution. What happened? Describe the site distribution. Mastodons are known to have been browsers, eating leaves and woody plant tissue. What can we say about the kinds of habitats or environments that mastodons may have occupied?

Figure 4. The search tools provide the opportunity to search for a single taxon or for multiple taxa simultaneously.

4 More Exploration with Explorer: Additional SERC Teaching Exercises

Additional exercises, focused on undergraduate, or senior high school students, are available on the SERC website (https://serc.carleton.edu/neotoma/activities.html). Other examples include: (1) Species distributions in response to environmental gradients in the Upper Midwest of the United States – an example using the Neotoma database is a unit for undergraduate students, focused on examining environmental gradients over short spatial scales and the ways these gradients can change over time; and (2) Climate Change and Mammal Dispersal, which is directed toward senior high school and early undergraduate students, helping them understand dispersal in small mammals through time in response to climate change.

Each of these exercises guides students through the process of accessing and exploring data from the Neotoma database to answer a scientific question or questions. This kind of pedagogical process can be referred to as scaffolding, or guided inquiry. Scaffolding provides a knowledge framework that allows students to discover their own answers while preventing the frustration they might experience if simply turned loose to answer the question (e.g., Hmelo-Silver et al., 2007).

5 Educational Modules for Climate Change, Paleoecology, and Biogeography

An additional series of seven modules was developed by vertebrate paleontologists at Penn State University for participants to learn about climate change, paleoecology, and biogeography using the Neotoma Paleoecology Database. These modules have also been posted to the SERC website for distribution (https://serc.carleton.edu/neotoma/activities.html). These modules are primarily focused on mammals but can easily be adapted to other organisms such as beetles, plants, ostracodes, and other groups with rich data holdings in Neotoma. There are two basic module types: (1) Modules that provide background information about climate change, paleoecology, and biogeography (background modules 1–4), and (2) modules in which learners apply knowledge from the set of background modules to examine hypotheses using data taken from the Neotoma database (applied modules 5–7).

Background modules can be used to introduce learners to fundamental concepts in climate change, paleoecology, and biogeography. For instance, in the paleoecology background module (Module 2), relationships among climate variables (moisture and temperature) are discussed and illustrated (e.g., the correspondence between latitude and temperature and correlation between longitude and precipitation in the eastern United States). Relationships between the modern distributions of mammals within the Neotoma database are compared to climate variables. From this, students can see that eastern and western limits of many species' distributions in the eastern United States are limited by water availability, whereas northern and southern range limits are temperature-dependent. The paleoecology background module (Module 2) also briefly discusses underlying assumptions linking climate variables and species distributions and how additional factors can come into play. For instance, the distribution of the prairie dog (*Cynomys ludovicianus*) may be controlled by climate but also by soil properties such as depth, grain size, and moisture. A series of questions reinforce the learning objectives at the end of each module. The modules also provide links to other web pages and literature that provide further explanation of the principles.

In the applied modules, learners build on the knowledge they have gained from the background modules and apply it to analyses of Neotoma data. Each module begins with a statement about the process to be learned, and an example of similar analysis is provided. Learners develop a hypothesis (e.g., "If the climate warms, then a species distribution will expand westward"). The student can then test the hypothesis, by examining changes in the distribution of the species during different times of climate change using climate data from the Greenland Ice Core Project (GRIP: NGRIP members, 2004), covering the transition from full glacial to late glacial (~20,000 to 13,000 years ago) or late glacial to late Holocene (~13,000 years ago to present). Mammal species are assigned to students as they continue the exercise. If the hypothesis is false, then the student is asked to think about why it failed. The student is then taken to a file that explains what potentially happened with their analysis. Finally, on the basis of these analyses, the student is asked to draw lessons from the past, thinking about the potential response of these species to current climate trends.

For biogeographic analyses, students are asked to examine changes in species distributions through time. Various exercises then focus on when a species arrived in an area and how long it persists. With the addition of FAUNMAP II and MIOMAP data that are currently being uploaded, participants will be able not only to work with the glacial-to-interglacial warming of the late Quaternary, but also to go back through the Miocene (25 million years ago). The Miocene time period spans a long-term cooling in the Earth system, the onset of ice sheets in the Northern Hemisphere, and regionally increasing aridity. These exercises will involve rates of immigration and extinction for various mammal species.

We are developing additional modules with an emphasis on the emerging field of conservation paleobiology (Dietl and Flessa, 2011; Dietl et al., 2015; Barnosky et al., 2017) and more sophisticated analyses of groups of species and past community dynamics.

6 Programmatic Access for Macro-scale Paleoecological Research (APIs, neotoma R, and GitHub)

For advanced undergraduates and graduate students, we have created educational resources to introduce them to the programmatic tools for accessing Neotoma data. Neotoma Explorer is excellent for quick-look visualizations, data exploration, and introductory teaching, but Neotoma's programmatic interfaces are better suited for large-scale data downloads and analyses. Neotoma supports programmatic data access through an API (https://api.neotomadb.org/) that provides users with the ability to access data from Neotoma using programming languages such as JavaScript, R, or Python. The neotoma R package (https://github.com/ropensci/neotoma; Goring et al., 2015) offers a convenient system for exporting Neotoma data into the R environment, which is widely used by ecologists and paleontologists for statistical data analysis. The R package is documented with worked examples drawn from the literature to showcase the ways that the R package can be used in research workflows (Goring et al., 2015).

Over the past several years, we and others have led several short courses and workshops to show early career scientists how to use the neotoma R package and APIs to obtain data from Neotoma and pass it to their analytical workflows, with workshops at the Society for Vertebrate Paleontology, International Biogeography Society, American Quaternary Association, and other venues. As part of these workshops, we have developed training materials that are hosted on GitHub (http://neotomadb.github.io/workbooks.html). GitHub (http://github.com) is a particularly useful system for open-source software development because of its ability to host and version programmatic code that can be accessed and modified by multiple developers. GitHub can also host public webpages (using GitHub Pages; https://pages.github.com/), providing the opportunity to host educational material, along with the source code for programmatic workflows. Our educational and training workshops for graduate students and early career scientists make frequent use of GitHub pages and much material can be found there (http://neotomadb.github.io/workbooks.html). All materials are posted with an MIT license, so any interested student or educator can access workshop resources, download copies, and modify them for their own research or educational purposes.

Teaching materials on GitHub use the software principle of version control to borrow elements across workshops, while allowing workshop leaders to adopt best practices from core workshop materials. This provides the ability to both reuse well-established teaching materials and to tailor them for new audiences. Workshop material for five separate workshops is currently available in the Neotoma Workshops GitHub repository. For example, materials from the 2016 SVP meeting are posted there (https://neotomadb.github.io/workbooks/Workshop_SVP2016_v0.1.html#the-neotoma-package), and new workshop materials are being actively developed.

6.1 Model Workflows: Chronology Construction and Climate Reconstruction

Two of the workflows posted to the Neotoma workbooks page (http://neotomadb.github.io/workbooks/AgeModels.html) were not produced for specific workshops but instead were designed as model workflows for common tasks in paleoecology and paleoclimatology: chronology construction (Fig. 5) and paleoclimatic reconstruction via transfer functions. Both are complex tasks. Chronology construction is central to making age

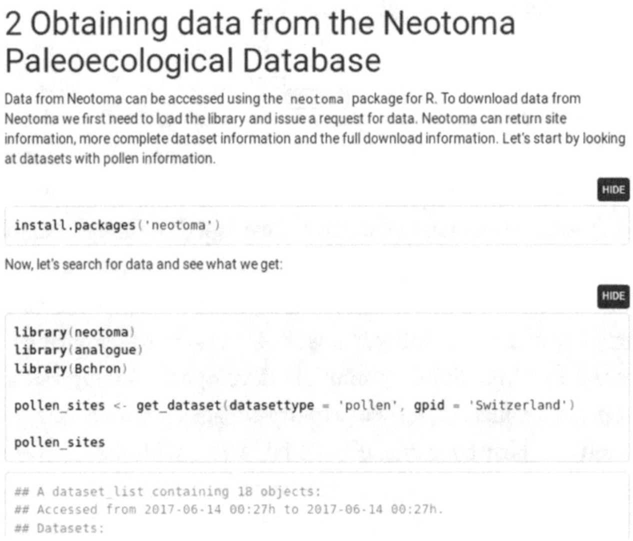

2 Obtaining data from the Neotoma Paleoecological Database

Data from Neotoma can be accessed using the `neotoma` package for R. To download data from Neotoma we first need to load the library and issue a request for data. Neotoma can return site information, more complete dataset information and the full download information. Let's start by looking at datasets with pollen information.

HIDE

```
install.packages('neotoma')
```

Now, let's search for data and see what we get:

HIDE

```
library(neotoma)
library(analogue)
library(Bchron)

pollen_sites <- get_dataset(datasettype = 'pollen', gpid = 'Switzerland')

pollen_sites
```

```
## A dataset_list containing 18 objects:
## Accessed from 2017-06-14 00:27h to 2017-06-14 00:27h.
## Datasets:
```

Figure 5. Educational material presented on Neotoma's GitHub pages show the initial stages of constructing chronologies for paleoecological records.

inferences and placing geological events in a temporal framework (Grimm et al., 2014; Harrison et al., 2016), while transfer functions and paleoecological data are often used to reconstruct past climates (e.g., Bartlein et al., 2011; Viau et al., 2012; Marsicek et al., 2018), but fraught with methodological issues (Juggins, 2013). Within the Quaternary geosciences, there are many techniques that can be used to construct chronologies, and packages that can be used to reconstruct climate from proxy data. Age-depth models such as *Bacon* (Blaauw and Christen, 2011) and *BChron* (Haslett and Parnell, 2016) for chronologies, or *rioja* (Juggins, 2017) and *analogue* (Simpson, 2007) for climate reconstruction, have a number of settings and requirements. However, the paleoecological community is highly distributed among institutions, and few departments or institutions offer advanced courses in quantitative paleoecological methods, so many students might not encounter a course on chronology construction or paleoclimate reconstruction. Hence, the educational materials hosted on Neotoma's GitHub Pages site (Fig. 5) offer example code and narrative discussion of a best practices workflow process in an approachable format, for both chronology construction (http://neotomadb.github.io/workbooks/AgeModels.html) and paleoclimatic reconstruction with fossil pollen records (https://neotomadb.github.io/workbooks/clim_notebook.html). These can be used in a lab setting for Earth science educators, or for students (graduate or senior undergraduates) who wish to undertake similar analysis.

7 Flyover Country®: Beyond the Classroom

Creating open-access, high-quality community-curated scientific data resources such as Neotoma opens up novel opportunities to engage with entirely new audiences. One example is the use of Neotoma data in the Flyover Country® mobile app for geoscience (supported by the National Science Foundation; Fig. 6), available for iOS and Android devices. The app is designed to be used by travelers (or armchair explorers) to learn more about the geological world around them as they travel over it. Flyover Country® draws upon multiple resources, including Macrostrat, Paleobiology Database, and Neotoma. For Neotoma, Flyover Country® uses the Neotoma API to find fossil occurrences of extinct "charismatic megafauna" (mastodons, saber-tooth tigers, giant beavers, etc.) along a traveler's path. Users click on the map to establish their travel waypoints, and then Flyover Country® submits a spatial query directly to Neotoma. The API structure allows Flyover Country® to link the taxon name (e.g.,

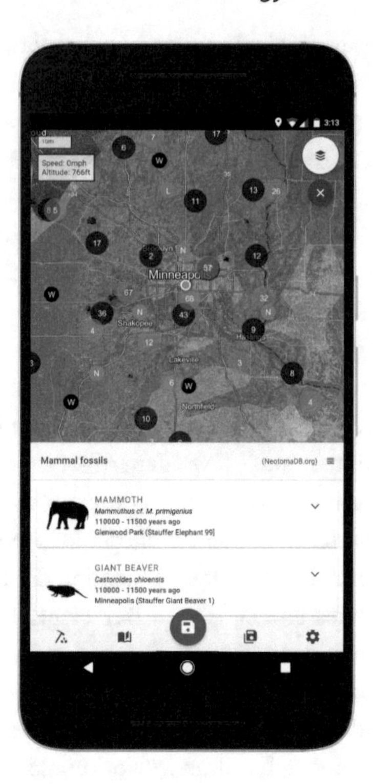

Figure 6. Flyover Country® provides a platform to access and engage with Earth science data from a mobile device. Flyover Country® is connected to databases such as Neotoma though the use of Neotoma's API, and adds another layer of usability to the raw data resources of Neotoma.

Mammut) with the Wikipedia article about that animal, which is then also displayed in the app, enriching the learning experience of the user. Mobile and web app development is within the reach of undergraduate courses, and Neotoma's well-documented API and the ready availability of Neotoma data through the API means that those apps can be easily populated with research-quality scientific data. Neotoma's spatial, multivariate, and temporal data provide additional interesting opportunities for data visualization in such courses.

8 Discussion

8.1 CCDRs at the Research-Outreach Nexus

Community-curated data resources (CCDRs; Williams et al. 2018) are managed resources whereby data are contributed by a disciplinary community and subsequently managed and curated by the same community. They are often motivated by research objectives, by a community of researchers attempting to answer macro-scale research questions that can only be answered by analyzing many individual site-level datasets with some standard data representation. Traditionally, CCDRs have been a resource for experts only, serving systems of record and as vehicles for collating, aggregating, cleaning, and analyzing data (Kapoor et al., 2015). As mechanisms to access, share, and visualize data improve across the geosciences, these community-curated research data resources generate new opportunities for individuals and groups outside of the traditional research community to access and interact with scientific data. This in turn moves these databases from systems of record and research to systems of engagement, opening up new social and collaborative elements to these platforms. As such, building educational activities that closely interlink with CCDRs represents a major opportunity for broader engagement, both within dispersed networks of scientific researchers and out to new and previously unreached communities.

8.2 The Role for CCDRs in Outreach

By acting as a portal for outreach, CCDRs can serve multiple, overlapping communities, and can bring them together in collaborative endeavors (Fig. 7). For example, the project "Neotoma: Community-led Cyberinfrastructure for Global Change Research" (NSF-Geoinformatics), in partnership with SERC, was able to bring together a broad group of researchers and Earth science educators, connecting and leveraging their diverse expertises (Fig. 7, Database, Educators, Data Contributors and Data Accessors). The strength of these curricular materials is multifaceted: the exercises make use of the rich data available through the Neotoma database, as well as its data visualization tools; they engage students in authentic data inquiry to answer questions of interest to scientists; and they illustrate the use of paleoecological data to understand the impacts of climate change processes happening today. Tapping the expertise of these communities – paleoecologists and educational researchers – produced more robust educational materials than either community would have produced alone. Thus, CCDRs provide an opportunity to link researchers to educators, and educators to the broad range of cutting-edge research activities, building on the strengths of each group. The network in Figure 7 is stronger than any of the set of parts on their own.

The ongoing informatics revolution, the rise of data sciences as a distinct field of inquiry (Blei and Smyth, 2017), and the associated emergence of paleoecoinformatics (Brewer et al., 2012) has enhanced the paleoecology community's suite of tools, bringing geospatial tools and the technical

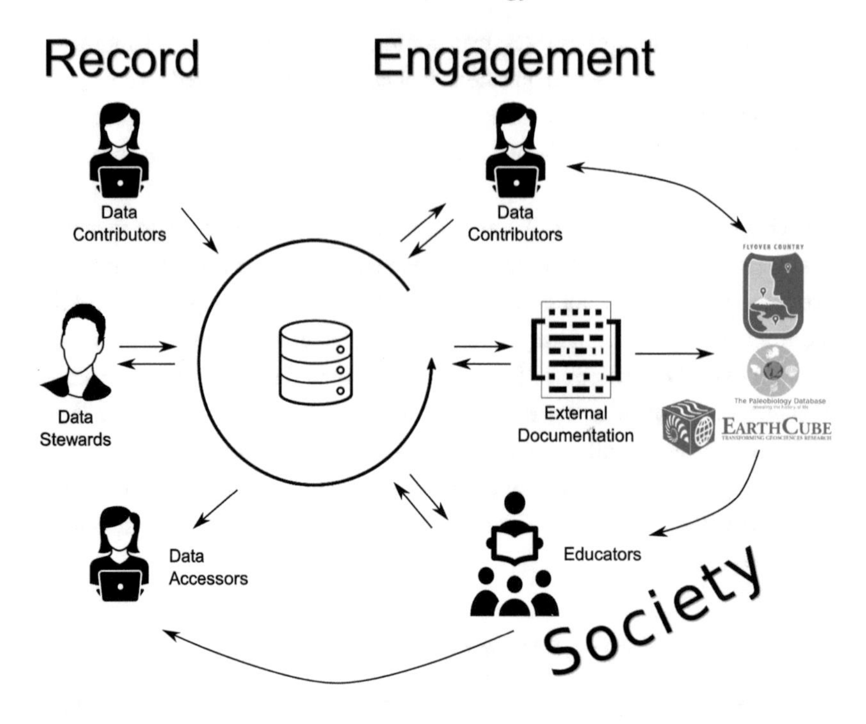

Figure 7. CCDRs involved in engagement can access new communities outside of the traditional sphere of activity in the research community. Here various activities allow a database to engage with community experts (data stewards), data contributors, researchers accessing data, but also educators and individuals involved in outreach activities through platforms such as Flyover Country® by providing well-documented workflows for accessing and understanding the data within the database. Clip art of the woman at a computer, created by Nikita Kozin and clip art of the database by Vicons Design, both from the Noun Project (http://nounproject.com), licensed under a CC-0 license.

capacity to combine web or mobile mapping tools with data resources. There is no one-size-fits-all solution to finding and accessing data and using these data to understand core concepts; different pathways serve different student and user populations. The development of the Neotoma Explorer (http://apps.neotomadb.org/explorer) provides an interface for secondary school students and beyond, while access through the Neotoma API and the neotoma R package (Goring et al., 2015) allows the integration of paleoecological data into post-secondary curriculum. With the emergence of larger cyberinfrastructure programs, such as NSF's EarthCube program, there will be further opportunities to introduce technology as a core component of curriculum design within the geosciences. CCDRs act as important translational actors in this ecosystem, drawing on best practices in both the data and educational sciences, and applying them to scientific outreach programs with the assistance of education experts (Williams et al., 2018).

In summary, there are a number of educational opportunities available online for the Neotoma Paleoecology Database, with more being actively developed. All are open source and can be modified to explore different questions or data types (e.g., plants, mammals, diatoms, or limnological datasets). They serve a broad range of audiences, from secondary to undergraduate to graduate-level students. CCDRs thus can serve dual missions, both as critical research infrastructure pedagogical infrastructure, supporting distributed research communities, and as opportunities for education and outreach. We welcome further contributions.

References

Araújo, M. B., Nogués-Bravo, D., Diniz-Filho, J. A. F., Haywood, A. M., Valdes, P. J., & Rahbek, C. (2008). Quaternary climate changes explain diversity among reptiles and amphibians. *Ecography*, **31**, 8–15.

Barnosky, A. D., Hadly, E. A., Gonzalez, P., Head, J., Polly, P. D., Lawing, A. M., Eronen, J. T., Ackerly, D. D., Alex, K., Biber, E., Blois, J., Brashares, J., Ceballos, G., Davis, E., Dietl, G. P., Dirzo, R., Doremus, H., Fortelius, M., Greene, H. W., Hellmann, J., Hickler, T., Jackson, S. T., Kemp, M., Koch, P. L., Kremen, C., Lindsey, E. L., Looy, C., Marshall, C. R., Mendenhall, C., Mulch, A., Mychajliw, A. M., Nowak, C., Ramakrishnan, U., Schnitzler, J., Shrestha, K. D., Solari, K., Stegner, L., Stegner, M. A., Stenseth, N. C., Wake, M. H., & Zhang, Z. (2017). Merging paleobiology with conservation biology to guide the future of terrestrial ecosystems. *Science*, **355**, eaah4787.

Bartlein, P., Harrison, S., Brewer, S., Connor, S., Davis, B., Gajewski, K., Guiot, J., Harrison-Prentice, T., Henderson, A., Peyron, O., Prentice, I. C., Scholze, M., Seppä, H., Shuman, B., Sugita, S., Thompson R. S., Viau, A. E., Williams, J., & Wu, H. (2011). Pollen-based continental climate reconstructions at 6 and 21 ka: A global synthesis. *Climate Dynamics*, **37**, 775–802.

Blaauw, M., & Christen, J. A. (2011). Flexible paleoclimate age-depth models using an autoregressive gamma process. *Bayesian Analysis*, **6**, 457–474.

Blei, D. M., & Smyth, P. (2017). Science and data science. *Proceedings of the National Academy of Sciences*, **114**, 8689–8692.

Blois, J. L., Feranec, R. S., & Hadly, E. A. (2008). Environmental influences on spatial and temporal patterns of body-size variation in California ground squirrels (*Spermophilus beecheyi*). *Journal of Biogeography*, **35**, 602–613.

Blois, J. L., McGuire, J. L., & Hadly, E. A. (2010). Small mammal diversity loss in response to late-Pleistocene climatic change. *Nature*, **465**, 771.

Brewer, S., Jackson, S. T., & Williams, J. W. (2012). Paleoecoinformatics: Applying geohistorical data to ecological questions. *Trends in Ecology & Evolution*, **27**, 104–112.

Davis, M. B., & Shaw, R. G., (2001). Range shifts and adaptive responses to Quaternary climate change. *Science*, **292**, 673–679.

Dietl, G. P., & Flessa, K. W. (2011). Conservation paleobiology: Putting the dead to work. *Trends in Ecology & Evolution*, **26**, 30–37.

Dietl, G. P., Kidwell, S. M., Brenner, M., Burney, D. A., Flessa, K. W., Jackson, S. T., & Koch, P. L. (2015). Conservation paleobiology: Leveraging knowledge of the past to inform conservation and restoration. *Annual Review of Earth and Planetary Sciences*, **43**, 79–103.

Freeman, S., Eddy, S. L., McDonough, M., Smith, M. K., Okoroafor, N., Jordt, H., & Wenderoth, M. P. (2014). Active learning increases student performance in science, engineering, and mathematics. *Proceedings of the National Academy of Sciences*, **111**, 8410–8415.

Gill, J. L., Williams, J. W., Jackson, S. T., Lininger, K. B., & Robinson, G. S. (2009). Pleistocene megafaunal collapse, novel plant communities, and enhanced fire regimes in North America. *Science*, **326**, 1100–1103.

Goring, S., Dawson, A., Simpson, G., Ram, K., Graham, R., Grimm, E., & Williams, J. (2015). neotoma: A programmatic interface to the Neotoma Paleoecological Database. *Open Quaternary*, **1**.

Grimm, E. (2008). Neotoma: An ecosystem database for the Pliocene, Pleistocene, and Holocene. *Illinois State Museum Scientific Papers E Series*, **1**.

Grimm, E. C., Blaauw, M., Buck, C., & Williams, J. W. (2014). Age models, chronologies, and databases workshop: Complete report and recommendations. *PAGES Magazine*, **22**, 104.

Harrison, M., Baldwin, S., Caffee, M., Gehrels, G., Schoene, B., Shuster, D., & Singer, B. (2016). Geochronology: It's about time. *Eos*, **97**, 12–13.

Haslett, J., & Parnell, A. C. (2008). A simple monotone process with application to radiocarbon-dated depth chronologies. *Journal of the Royal Statistical Society: Series C (Applied Statistics)*, 57(4), 399–418.

Hmelo-Silver, C. E., Duncan, R. G., & Chinn, C. A. (2007). Scaffolding and achievement in problem-based and inquiry learning: A response to Kirschner, Sweller, and Clark (2006). *Educational Psychologist*, **42**, 99–107.

Hunter, A.-B., Laursen, S. L., & Seymour, E. (2007). Becoming a scientist: The role of undergraduate research in students' cognitive, personal, and professional development. *Science Education*, **91**, 36–74.

Juggins, S. (2013). Quantitative reconstructions in palaeolimnology: New paradigm or sick science? *Quaternary Science Reviews*, **64**, 20–32.

(2017). rioja: Analysis of Quaternary Science Data, R package version (0.9–15.1). (http://cran.r-project.org/package=rioja), accessed 15 June 2018.

Kapoor, S., Mojsilovic, A., Strattner, J. N., & Varshney, K. R. (2015). From open data ecosystems to systems of innovation: A journey to realize the promise of open data, in Bloomberg Data for Good Exchange Conference, https://pdfs.semanticscholar.org/ebfa/cdfc9da14c5b54791e6fba89bd7a6c7809d0.pdf, accessed 1 August 2018.

Latałowa, M., & Knaap, W. O. van der (2006). Late Quaternary expansion of Norway spruce Picea abies (l.) Karst in Europe according to pollen data. *Quaternary Science Reviews*, **25**, 2780–2805.

Laursen, S., Hunter, A.-B., Seymour, E., Thiry, H., & Melton, G. (2010). *Undergraduate research in the sciences: Engaging students in real science*. San Francisco: John Wiley & Sons.

Lopatto, D. (2010). Undergraduate research as a high-impact student experience. *Peer Review*, **12**, 27.

Manduca, C., & Mogk, D. (2002). Using data in undergraduate science classrooms, in Final Report of the National Science Digital Library Workshop, https://d32ogoqmya1dw8.cloudfront.net/files/usingdata/UsingData.pdf, accessed 1 August 2018.

Marsicek, J., Shuman, B. N., Bartlein, P. J., Shafer, S. L., & Brewer, S. (2018). Reconciling divergent trends and millennial variations in Holocene temperatures. *Nature*, **554**, 92.

Masson-Delmotte, V., Schulz, M., Abe-Ouchi, A., Beer, J., Ganopolski, A., Rouco, J. G., Jansen, E., Lambeck, K., Luterbacher, J., Naish, T., Osborn, T., Otto-Bliesner, B., Quinn, T., Ramesh, R., Rojas, M., Shao, X., & Timmermann, A. (2013). Information from paleoclimate archives. In *Climate Change: The Physical Science Basis. Contribution of Working Group I to the Fifth Assessment Report of the Intergovernmental Panel on Climate Change*. Cambridge University Press.

Mathewes, R. W., & Heusser, L. E. (1981). A 12 000 year palynological record of temperature and precipitation trends in southwestern British Columbia. *Canadian Journal of Botany*, **59**, 707–710.

Mayewski, P. A., Rohling, E. E., Stager, J. C., Karlén, W., Maasch, K. A., Meeker, L. D., Meyerson, E. A., Gasse, F., Kreveld, S. van, Holmgren, K., Lee-Thorp, J., Rosqvist, Rack, G. F., Staubwasser, M., Schneider, R. R., & Steig, E. J. (2004). Holocene climate variability. *Quaternary Research*, **62**, 243–255.

North Greenland Ice Core Project (NGRIP) members (2004). High-resolution record of Northern Hemisphere climate extending into the last interglacial period. *Nature*, **431**, 147–151.

Ordonez, A., & Svenning, J.-C. (2017). Consistent role of Quaternary climate change in shaping current plant functional diversity patterns across European plant orders. *Scientific Reports*, **7**, 42988.

Resnick, I., Atit, K., & Shipley, T. (2012). Teaching geologic events to understand geologic time. In K. A. Kastens and C. A. Manduca, eds., *Earth and Mind II: A Synthesis of Research on Thinking and Learning in the Geosciences*. Boulder, CO: Geological Society of America, pp. 41–43.

Resnick, I., Shipley, T. F., Newcombe, N., Massey, C., & Wills, T. (2011). Progressive alignment of geologic time, http://w3w.spatiallearning.org/archives/showcase_archive/showcase_pdfs/showcase_resnick_sep2011.pdf, accessed 1 August 2018.

Simpson, G. L. (2007). Analogue methods in palaeoecology: Using the analogue package. *Journal of Statistical Software*, **22**, 1–29.

Viau, A., Ladd, M., & Gajewski, K. (2012). The climate of North America during the past 2000 years reconstructed from pollen data. *Global and Planetary Change*, **84**, 75–83.

Wanner, H., Beer, J., Bütikofer, J., Crowley, T. J., Cubasch, U., Flückiger, J., Goosse, H., Grosjean, M., Joos, F., Kaplan, J. O., Küttel, M., Müllerg, S. A., Prentice, I. C., Solomina, O., Stocker, T. F., Tarasov, P., Wagner, M., & Widmannm, M. (2008). Mid-to late Holocene climate change: An overview. *Quaternary Science Reviews*, **27**, 1791–1828.

Williams, J. W., Grimm, E. C., Blois, J., Charles, D. F., Davis, E., Goring, S. J., Graham, R. W., Smith, A. J., Anderson, M., Arroyo-Cabrales, J., Ashworth, A. C., Betancourt, J. L., Bills, B. W., Booth, R. K., Buckland, P. I., Curry, B. B., Giesecke, T., Jackson, S. T., Latorre, C., Nichols, J., Purdum, T., Roth, R. E., Stryker, M., & Takahara, H. (2018). The Neotoma Paleoecology Database: A multi-proxy, international community-curated data resource. *Quaternary Research*, **89**, 156–177.

Acknowledgments and Authorship

Authorship is ordered by lead author (SJG) and subsequently by alphabetical order. SJG produced the main text and figures. Co-authors provided text and editing, and assisted in the workshop development and module creation. The authors would like to acknowledge contributors to the Neotoma Paleoecology Database and Neotoma data stewards. Work on this paper was supported by NSF Awards NSF-1541002, NSF-1550707 and NSF-1550707.

Cambridge Elements

Elements of Paleontology

About the Series

The Elements of Paleontology series is a publishing collaboration between the Paleontological Society and Cambridge University Press. The series covers the full spectrum of topics in paleontology and paleobiology, and related topics in the Earth and life sciences of interest to students and researchers of paleontology.

The Paleontological Society is an international nonprofit organization devoted exclusively to the science of paleontology: invertebrate and vertebrate paleontology, micropaleontology, and paleobotany. The Society's mission is to advance the study of the fossil record through scientific research, education, and advocacy. Its vision is to be a leading global advocate for understanding life's history and evolution. The Society has several membership categories, including regular, amateur/avocational, student, and retired. Members, representing some 40 countries, include professional paleontologists, academicians, science editors, Earth science teachers, museum specialists, undergraduate and graduate students, postdoctoral scholars, and amateur/avocational paleontologists.

Elements of Paleontology

Elements in the Series

These Elements are contributions to the Paleontological Short Course on *Pedagogy and Technology in the Modern Paleontology Classroom* (organized by Phoebe Cohen, Rowan Lockwood, and Lisa Boush), convened at the Geological Society of America Annual Meeting in November 2018 (Indianapolis, Indiana USA).

Flipping the Paleontology Classroom: Benefits, Challenges, and Strategies
Matthew E. Clapham

Integrating Macrostrat and Rockd into Undergraduate Earth Science Teaching
Phoebe A. Cohen, Rowan Lockwood, and Shanan Peters

Student-Centered Teaching in Paleontology and Geoscience Classrooms
Robyn Mieko Dahl

Beyond Hands On: Incorporating Kinesthetic Learning in an Undergraduate Paleontology Class
David W. Goldsmith

Incorporating Research into Undergraduate Paleontology Courses: Or a Tale of 23,276 Mulinia
Patricia H. Kelley

Utilizing the Paleobiology Database to Provide Educational Opportunities for Undergraduates
Rowan Lockwood, Phoebe A. Cohen, Mark D. Uhen, and Katherine Ryker

Integrating Active Learning into Paleontology Classes
Alison N. Olcott

Dinosaurs: A Catalyst for Critical Thought
Darrin Pagnac

Confronting Prior Conceptions in Paleontology Courses
Margaret M. Yacobucci

The Neotoma Paleoecology Database: A Research-Outreach Nexus
Simon J. Goring, Russell Graham, Shane Oeffler, Amy Myrbo, James S. Oliver, Carol Ormond, and John W. Williams

Equity, Culture, and Place in Teaching Paleontology: Student-Centered Pedagogy for Broadening Participation
Christy C. Visaggi

A full series listing is available at: www.cambridge.org/EPLY

Printed in the United States
By Bookmasters